YOUR KNOWLEDGE HAS VALUE

AF150729

- We will publish your bachelor's and master's thesis, essays and papers

- Your own eBook and book - sold worldwide in all relevant shops

- Earn money with each sale

Upload your text at www.GRIN.com and publish for free

Roman Kühn

Anthropogenic influence on the Tibetan Plateau

GRIN Verlag

Bibliografische Information der Deutschen Nationalbibliothek:

Die Deutsche Bibliothek verzeichnet diese Publikation in der Deutschen National-
bibliografie; detaillierte bibliografische Daten sind im Internet über http://dnb.d-
nb.de/ abrufbar.

Imprint:

Copyright © 2014 GRIN Verlag GmbH
Druck und Bindung: Books on Demand GmbH, Norderstedt Germany
ISBN: 978-3-656-72417-9

This book at GRIN:

http://www.grin.com/en/e-book/279211/anthropogenic-influence-on-the-tibetan-
plateau

GRIN - Your knowledge has value

Der GRIN Verlag publiziert seit 1998 wissenschaftliche Arbeiten von Studenten, Hochschullehrern und anderen Akademikern als eBook und gedrucktes Buch. Die Verlagswebsite www.grin.com ist die ideale Plattform zur Veröffentlichung von Hausarbeiten, Abschlussarbeiten, wissenschaftlichen Aufsätzen, Dissertationen und Fachbüchern.

Visit us on the internet:

http://www.grin.com/

http://www.facebook.com/grincom

http://www.twitter.com/grin_com

RWTH Aachen
Geographisches Institut
Großes Regionalpraktikum: China
Sommersemester 2014

10.08.2014

Anthropogenic influence on the Tibetan Plateau

Roman Kühn

6. Semester
Studienfach: B.Sc. Angewandte Geographie

Table of content

1. Introduction

The Tibetan Plateau is probably the most important and one of the most fragile areas in the world. With its enormous tectonic geomorphology forms and its unique ecosystem the Tibetan Plateau has an huge impact on the asian continent and even on the whole global climate (Cui/Graf 2009:49). For hundred of years this area was inhabited by few nomads, who practiced a sustainable approach of life in this space. Initiating with the reforms of " the Great Leap Forward, the Opening Door Policy and the Globalisation" an imigration flow began to discover and settle down. Those human activities had an influence on the climate change and land-cover change. The consequences resulting thereoff are evident and measurable and the aftereffects of that development cant yet be foreseen in their whole scale.

The purpose of this study is to examinethe anthropogenic influence on the Tibetan Plateau in recent years. The structure applied to this paper begins with a geographic classification of the Tibetan Plateau with its geographical, geomorphological and climatical charecteristics. The second part discribes the anthropogenic influence on the Plateau area and its demographical and agricltural development since the 1950s. The impacts of this influence on the vegetational and soil change are being discribed in the third part and the conclusion finalises this study.

2. Geographic Classification of the Tibetan Plateau

The Tibetan Plateau is a "huge mountainous area of the Eurasian continent with an average altitude of 4,000m" above sea level (Cui,Graf 2008:48). It is often recalled the "Roof of the World" and it accomodates the "(...) largest glaciated area outside the Polar Region(...)" and therefore "(...) plays an important role in regional and global atmospheric circulation." (Lehmkuhl/Schlütz 2009:1449). The Plateau stretches between the 28 and 40 degrees North Latitude and 70 to 110 degrees East Latitude. The territorium spreads for about 1280 km from north to south and 1600 km from east to west. Fig.1 describes the geographical and geomorphological characteristica of this area. It includes the Tibetan Autonomous Region, Qinghai province, parts of Sichuan and Gansu provinces, and parts of the Xinjiang Uygur Autonomous region (Chen et al. 2006:34). According to Lehmkuhl/Schlütz (2009:1449) "The Plateau and the bordering mountain ranges are influenced by five major climatc systems:

the mid-lattitude westerlies, the South and East Asian monsoons, the Siberian high-pressure system and the El Niño Southern Oscillation (ENSO)." Summertime consists the most precipitation on while winters are commonly cold and dry. The precipitation decreases from southeast to northwest and hence is responsible for a moisture gradient from humid to arid with an allocation of land covers from forest, grasland, shrubland, glaciers, meadow, desert and steppe (Cui 2005:7, Cui/Graf 2009:49). The Tibetan Plateau, or the so called Qinghai Xizang Plateau, is "the source region for several major rivers in southeastern and eastern Asia. Including the Yellow River, Yangtze River, Mekong River, and Salween River (Cui/Graf 2009:49). Therefore this region has an immense importance for the surrounding areas.

Fig. 1 Geographic Map of the Tibetan Plateau (Harris 2010:2)

3. Traditional Nomad pastoralism on the Tibetan Plateau

According to Harris (2010:1) "Livestock grazing is the dominant form of land use in arid biomes worldwide, and grazing lands of the Qinghai-Tibetan plateau." and therefore plays an important role for the landscape development and the transformation to a domesticated ecosystem (Lehmkuhl/Schlütz 2009:1450). The climatic circumstances of the grasslands with its sparse flora and its huge expanse of land are suitable for pastoral farming. The extensive livestock farming is being performed for thousands of years by the Nomadic inhabitants of the Tibetan Plateau (Miller 2007:2). Lehmkuhl/Schlütz state (2009:1450) that the anthropogenic influence of the

3

Titbetan nomadic culture changed the natural grazing system, inhabited by only wild herbivores, to an anthropozoogenic system. Over the past centuries the residents enveloped a sustainable approach for range and livestock farming under those difficult circumstances. According to Miller (2007:3) the Tibetan nomads graze their cattles at altitudes from 3,000 to 5,000 meters in environmental conditions that are too cold for crop cultivation and therefore dont compete with other agricultural pursuits. According to (Lehmkuhl/Schlütz 2009:1455) the pasturing has a significant influence on the "composition and structure of the vegetation.", resulting in a change of the vegetation types through the grazing intensity. Plants species like Anemone, Bistorta and Poaceae are suppressed and overcome by species like Leontopodium, Saussurea and Ligularia, which are more grazing resistent, in due to an increasing grazing pressure and therefore change the natural Biomass diversity through nomadical induced land-use. According to Lehmkuhl/Schlütz (2009:1468) the biomass diversity change in due to the transformation of the plants species have important influence in the local and regional climate and therefore "it is highly likely, that the anthropo-zoogenic vegetation hanges have affected synoptic conditions."

The economic growth of the VR China, during the intensive phase of mobilization, the Great Leap (1958/59) and the Opening Door Policy (1978), caused interventions in the traditional system of rotation, the transformation of grasslands into crop land and the market-oriented organization of the livestock production (Shen et al. 2008:56). This led to a development driven by the planned economy of the VR China and an increasing demand for livestock and agricultural goods. In addition to that it put the traditional sustainable system under pressure to support the market with those kinds of products. "Since 1978, when China initiated economic reform and an open-door policy, rapid land use and land cover change has taken place in most of its territory." (Weng 2002 cit. in Shen et al. 2008:56). The effect of that was the tremendous increase in livestock numbers, as stated by Du et al. (2004 cit. in Chen et al. 2006:35), "Over the last 30 years, livestock numbers across the TP have increased more than 200% due to inappropriate land management practice.", and the enormous impact on the ecological equilibrium wich had consequences on the climate system in addition (Lehmkuhl/Schlütz 2009:1450).

This overdevelopment of the already fragile ecosystem of the Tibetan Plateau leads to some extensive intrusion in the climatic change and degradation of soil and vegetation. One of the main problems of the livestock increase was the traditional farmers attitude (Bai et al. 2002 cit. in Harris 2010:7). Large numbers of farm animals are synonymous for the amount of wealth and financial securiy therefore large livestock numbers assure the farmers the possibility for unanticipated loss or outfalls of the productive livestock. Fig.2 illustraes the enormous increase in the livestock production of sheep and cattles in Tibet during 1978 and 1999. The amount of sheep increased for almost 110% and that of cattes for 250% by comparison to the reference year of 1978.

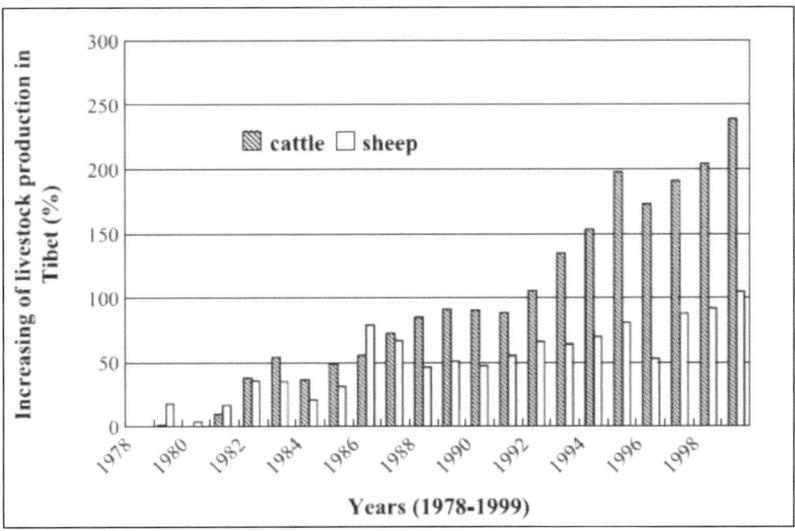

Fig. 2 Increase oflivestock prodction in Tibet (Du et al.2004:243)

Acompanied with the increase in the number of livestock and crops facilities and other state farms a migration to this rural area came along (Brown et al. 2008:228, Shen et al. 2008:57). Chinese farmers, who were not familiar with sustainabe farming on the Tibetan Plateau imigrated. The migration flows led to an increase of the conversion of grasslands to crop land and the cultivation of large amounts of land around the new settlements. According to Cui (2005:21) the "Population in the Autonomous Region has increased rapidly from 1960 (1.2 million) to 1990 (2.2 milion) (Chinese national statistics reports, contributing to both urbanization and

5

changing landscape." Fig.3 and Fig.4 illustrate the enormous increase in the demographic growth from 1952 (2.76 millions) up to 1997 (10.37 millions).

Year	Tibet Autonomous Region	Qinghai Province	Total
1952	115.00	161.38	276.38
1959	122.80	260.01	382.81
1960	126.98	248.65	375.63
1961	129.87	211.42	341.29

Fig.3 Demographic grow in Tibetan Plateau 1952-1961 (104 persons) (Fu 2001:73)

Year	1985	1990	1991	1993	1997
Tibet Autonomous Region	199.5	218.1	221.8	228.9	242.7
Qinghai Province	407.4	447.7	454.4	466.7	495.6
Ganzi Prefecture, Sichun Province	78.6	82.3	83.1	83.9	86.5
Aba Prefecture, Sichun Province	71.9	76.6	76.8	77.5	80.5
Muli county, Sichun Province	10.5	11.3	11.4	11.5	12.0
Gannan Prefecture, Gansu Province	51.4	58.2	55.9	60.7	64.3
Tianzhu county, Gansu Province	19.3	20.9	21.2	21.4	21.9
Diqing Prefecture, Yunnan Province	28.0	31.5	31.8	32.1	33.0
Tibetan Plateau	866.6	946.6	956.4	982.7	1036.5
Source: According to the statistical materials of each province and region					

Fig. 4 Demographic grow in Tibetan Plateau 1985-1987 (10^4 persons) (Fu 2001:74)

The need for labour force and a higher birth rate, due to less strict birth control (Harris 2010:7) around the plateau, were signifcant for this demographic trend (Gou et al. 2004:51)

"Along with the rapid growth of the population and the economic activities on the Qinghai-Xizang (Tibet) Plateau, the amount of the necessary meat, food, mineral ressources and fuel will increase. These have resulted in overgrazing, over cultivating, abusive mining practices for gold, over cutting and digging the vegetation as fuel, and over digging for herbs."

The increase in livestock production implicates the fact that there should be also an increase in the cultivated area, as seen in Fig.5, and "(...) an increase in the consumption of plant biomass from the grassland on the TP since there is no food import to the TP." (Chen et al. 2004:244). In addition to the human activities, like urbanization and demographic growth, this entails the unproportional use of the very fragile ecosystem anyway and causes the degradation of soil and harsh breaks in the climatic nature.this again leads to an advance of degradation of grasslands, and results in the loss of vegetation cover and deforestation, soil erosion and desertification.

	Total population (10⁴ persons)		Area of cultivate land (10³ hm²)	
	1952	1995	1952	1995
Tibet	115	240	163.3	222.1
Qinghai	146	481	464.6	589.9

Fig. 5 Population growth and conversion of cultivated land (Wang 1996 cit. in Fu 2001:80)

4. Soil degradation

The compound climatic and ecological circumstances caused by the altitude, causes the limitaion of the vegetation types and agricultural production capacities. The fragile soils are causing their own degradation (Harris 2010:4, Cui 2005:10). Additionaly anthropogenic influence burdens the Tibetan Plateau through the increase of population and agricultural locations and exacerbates the degradation of rangeland. According to Harris (2010:1) the main reason for the degradation is the over-stocking of livestock which leads to overgrazing. The livestock increase does not consider the limited carrying capacity of the rangelands and therefore the conditions can decline if more farm animals are kept than the rangelands can endure. Miller (2005, cit. in Harris 2010:3) stated that 2005 "there were an estimated 30 million sheep (including goats) and 12 million yaks on the QTP." According to Harris (2010:7) the massive amount of farm animals and the concentration of livestock near the residental areas affected the vegetation and the rangelands in a negative way. The construction of fences and therefore the absence of rotation entailed the intensity of pressure on them.

Fig.6 indicates the radical transformation of previous Grasland in the south-eastern and northern plateau area to cropland and to land used by livestocks due to inappropriate land management and economic pressure. The Tibetan rangelands are not suited for supporting an amount of this scale of livestock capacity and therefore the degradation can have enormous effects on the vegetation, soils and hydrology (Harris 2010:6).

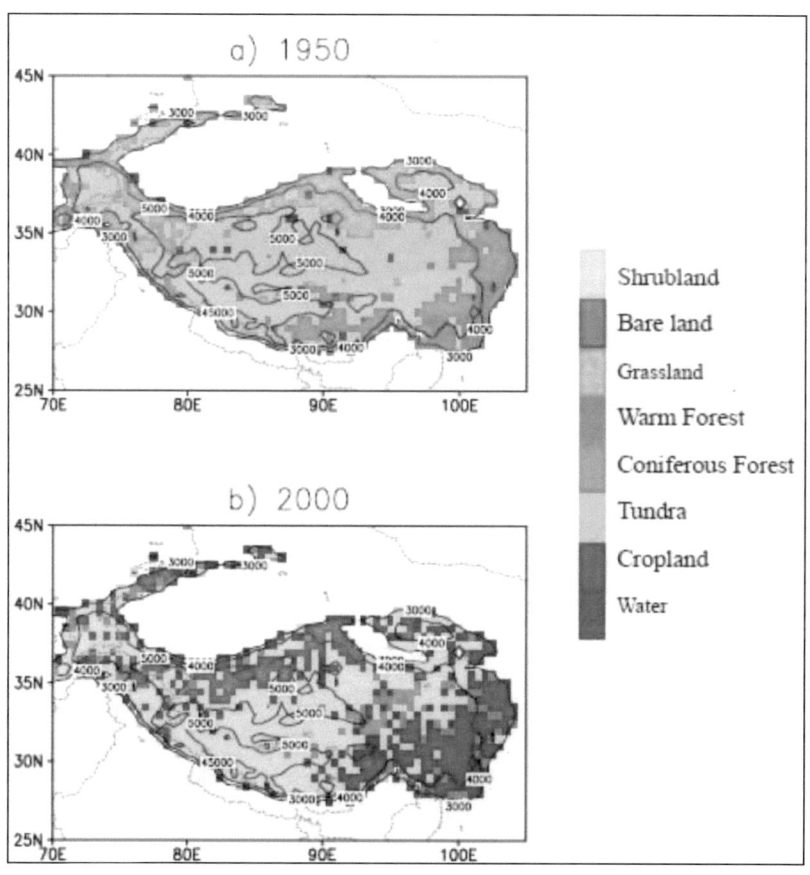

Fig.6 Land cover change on the Tibetan Plateau 1950(a)-2000(b) (Cui 2005:11)

4.1 Deforestation

The intensive pastural and agrarcultural land- use over the centuries resulted in a natural vegetation change. Through deforestation in the south-eastern part of the Tibetan Plateau the forest area was reduced from about 121×10^6 ha to 24×10^6 ha. According to Miehe et al. (2014:194) it is possible that the "current treeless and alpine pastures were forests that were converted to pastures to increase possible livestock population densities." and therefore the anthropogenic influence on the Plateau is the main driven influence on the current ecosystem. Miehe et al.

8

(2014:191) states that "(...) the Tibetan Plateau is not a natural environment but rather a human-induced replacement of grassland and forests (...)." The degradation of the hypothetical forest formations was achieved through deforestation and the usage of fire for an increase for the agricultultural land-use (Miehe et al. 2014:194,196). As stated by Miehe et al. (2014:204)

> "(...) and why the spruce forest was not able to re-establish in a forest climate can in our opinion most plausibly be explained by a scenario in which humans removed the spruce forest by fire to establish pastures and not to maintain forest."

Especially those tree formations were supressed in their regeneration with for humans and livestock accessile sites, "The degree of humas or livestock impact varied but was usually high. A closed cover of moss- an unfailing indicator of absence of disturbance-was very rarely found." (Miehe et al. 2014:202).

According to Tafel (1914, cit. in Lehmkuhl/Schlütz 2009:1452) the "Chinese deforestation and farming started at the end of the last century in the valleys and was intensified during the Cultural Revolution (1966-1976)." Cui (2005:12,52) states that since the 1950´s the forest area in the south-east almost vanished and has been substituted by cropland and built-up areas. According to Cui (2005:52-54) the deforestation accelerates soil erosion, the loss of biodeversity and "may impair forest functions of safeguarding watersheds and river flow." causing the possibility for catastrophic flooding.

The expansion of human activities implicated significant changes to the vegetation cover. According to Chen et al. (2004:247):

> "(...) it has changed the structure of plant communities and this structure changed not only restrained growth of grasses and sedges and contested against livestock for food, but also destroyed plant over and formed micro landforms, which induced loss of water and soil and essentialy degraded the grassland ecosystems."

Chen et al. (2004:247) and Harris (2010:5) also state that this loss in vegetation cover leads to a decrease in evapotranspiration, an increase of the albedo effect and therefore results in an increase in local temperature levels and amplifies soil erosion. Due to the increase of the local temperature gradient, as a result of change in vege-tation cover, the tundra area in the middle of the plateau vanished and transformed over the last 50 years to shrubland (Fig.6).

4.2 Soil erosion

Deforestation of forest areas and its transformation to crop-land and rangeland are the main reasons for the destroying of vegetation and soil layers and their degradation. The overgrazing, caused by the tremendous rise of livestock (Fig.2) leads to an increase of the performed pressure on the ecosystem and its soil. The overuse by the farm animals and agricultural measures leads to a devastation of soil structure resulting in the reduction of root area, the reduction of water storage capacity, the loss of vegetation cover and nutrient content (Fiedler 2001:441). As said by Miehe et al. (2014:205) the anthropogenic land-use therefore could be the main reason for the soil erosion since the early Holocene despite the sustainable land-use approach and the relatively limited number of livestock till the great expansion in the 20th century.

This problems results in a high degradation of soil and therefore high possibility for wind erosion especially in arid areas with lack of precipitation and that altitude like the Tibetan Plateau. Especially the fact that the Tibetan Plateau is " (...) one of the places with most strong-wind day-numbers in China. (...) the annual average wind speed is 3.0-4.0 m/s. (...) The time that the wind stronger than 17.2 m/s is 30-129 days(...)." (Gou 2004:49). This describes that the critical wind speed for uplifting sand is lasting for over 129 days and displays the enormous influence of aeolian erosion in the soil degradation and surface building processes. According to Miehe et al. (2014:196) soil erosion can also be a result of unsustainable use of mountainous terrain for agricultural use. Therefore the "Corresponding sediments are often colluvial silt and sand bodies on slopes and valley floors that bury older soil surfaces." and anttributable to the overstress can lead to slope erosion by runoff.

4.3 Desertification

The "United Nations Convention to Combat Desertification (UNCCD)" defined the term in the year 1994 as a "land degradation in arid, semi-arid and dry sub-humid areas resulting from various factors, including clmatic variations and human activities."

The desertification happening on the fragile ecosystem of the Tibetan Plateau is mostly determined by its own degradation caused by its arid climate conditions the lack of precipitation, the altitude and the low temperatures. The anthropogenic activities are additionally supporting this development of disertification by destroying the vegetation and its soil layers through overgrazing, deforestation and agricultural overuse (Gou et al. 2004:47). The animal husbandry and agroindustry exert intensive stress and destroy the soil structure with its rootssystems and its infiltration rate of percipitation. Large livestocks without the possibilty for rotation burden the ecosystem. The soils regenerative abilities decline under the stress due to the anthropogenic influence in terms of soil degradation (Rochette et al. 1989 cit. in Hammer 1999:274). The increase of evaporation and soil drain and the degredation of vegetational cover affect the desertification process. According to Gou (2004:47,48) the desertification is distributing around the Tibetan Plateau. The desert Area in the Qinghai province streched since the 1960s from 597×10^4 km^2 to almost 790×10^4 km^2 in the 1980s. This addition of desert land complies to 10.95% of the total provincial area.

5. Conclusion

The anthropogenic influence on the Tibetan Plateau exists for hundreds of years. The traditional and sustainable approach of local nomads was almost perfectly suited with its sparse and fragile ecosystem. Since the economic and social reforms of "The Great Leap Forward and the Opening Door Policy" of the chinese government the vast expanses of the mountainous regions of the Tibetan Plateau began to be intensively discovered and populated. A process of urbanization and utlization of former grasslands began. The rising demand for livestock and agricultural goods led to an tremendous increase of production facilities. This development conducted the overgrazing, soil degradation and finally to desertification. This changes are evident and measurable. But, as stated in the study before, the degradation of soil is not only the result of an anthropogenic influence but is also based on the fragile ecosystem of the Tibetan Plateau. It is just the fact that irresponsible economic activities led to this degradation. Fig.7 shows the Potential Land cover of the Tibetan Plateau and in comparison to the actual Land cover (Fig.6) its an evident change in the soil and vegetation structure. The future task of the chinese policy and inhabitants of the Tibetan Plateau who populated and cultivated the area is to secure this ecosystem and to grant sustainability in the aricultural business.

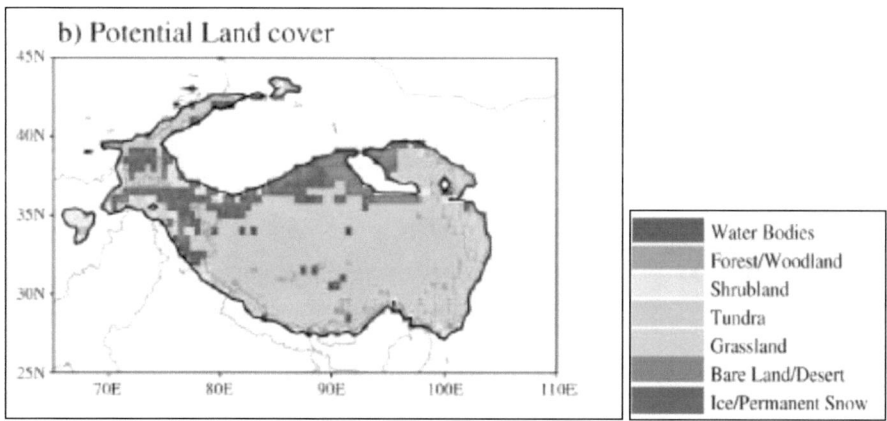

Fig. 7 Potential Land cover change (Chen et al. 2006:35)

6. References

Brown, C. G., Longworth, J. W., & Waldron, S. A. (2008): Sustainable Development in Western China: Managing People, Livestock and Grasslands in Pastoral Areas. Queensland, Australia: Edward Elgar Publishing.

Chen, S. et al. (2004): Mutual influence between human activities and climate change in the Tibetan Plateau during recent years. Global and Planetary Change 41 (2004), p. 241-249.

Chen, W. et al. (2006): Climate impacts of anthropogenic land use changes on theTibetan Plateau. Global and Planetary Change 54 (2006), p. 33-56.

Cui, X. (2005): Interactions between Climate and Land Cover Changes on the Tibetan Plateau. Hamburg.

Cui, X., & Graf, H.-F. (2009): Recent land cover changes on the Tibetan Plateau: a review. Climatic Change 94 (2009), p. 47-61.

Fiedler, H. J. (2001): Böden und Bodenfunktionen in Ökosystemen, Landschaften und Ballungsgebieten. Renningen: expert-verlag.

Fu, X. (2001): The Population changes and urban development. In Z. Du, Q. Zhang, & S. Wu, Mountain Geoecology and Sustainable Development of the Tibetan Plateau (GeoJournal Library) (2001), p. 71-88. Springer Verlag.

Gou, X. et al. (2004): Desertification and its relationship with permafrost degradation in Qinghai-Xizang (Tibet) plateau. Cold Regions Science and Technology 39 (2004), p. 47-53.

Hammer,T. (1999): Nachhaltige Entwicklung im Lebensraum Sahel - Ein Beitrag zur Strategietheorie nachhaltiger ländlicher Entwicklung. Münster: LIT.

Harris, R. (2010): Rangeland degradation on the Qinghai-TIbean plateau: A review of the evidence of its magnitude and causes. Journal of Arid Environments 74 (2010), p. 1-12.

Lehmkuhl, F.,Schlütz, F. (2009): Holocene climatic change and the nomadic Anthropocene in Eastern Tibet; palynological and geomorphological results from the Nianbaoyeze Mountains. Quaternary Reviews 28 (2009), p.1449-1471.

Miehe, G. et al. (2014): How old is the human footprint in the world`s largest alpine ecosystem? A review of multiproxy records from the Tibetan Plateau from the ecologists viewpoint. Quaternary Science Reviews 86 (2014), p.190-209.

Shen, Y., Wang, X., & Zheng, D. (2008): Land use change and its driving forces on the Tibetan Plateau during 1990-2000. Catena 72 (2008), p. 56-66.

UNCCD (1994): Desertification as agobal Problem
<http://www.ifad.org/pub/desert/scheda1.pdf> abgerufen am 04.05.2014